Leaving Certificate

BIOLOGY

(Ordinary and Higher Level)

Preamble

Policy Context

Science education in the senior cycle should reflect the changing needs of students and the growing significance of science for strategic development in Ireland.

Leaving Certificate Science syllabuses are designed to incorporate the following components:

- science for the enquiring mind or pure science, to include the principles, procedures and concepts of the subject as well as its cultural and historical aspects

- science for action or the applications of science and its interface with technology

- science that is concerned with issues–political, social and economic–of concern to citizens.

The three components are integrated within each science syllabus, with the first component having a 70% weighting. The remaining 30% should be allocated to the other two components in the ratio 3 to 1.

The science syllabuses, which are offered at two levels, Higher and Ordinary, will have approximately 180 hours of class contact time over a two-year period. They should be practically and experimentally based in their teaching.

• LEAVING CERTIFICATE BIOLOGY SYLLABUS •

CONTENTS

Introduction 2

**Unit One: Biology -
The Study of Life** 5

Unit Two: The Cell 15

Unit Three: The Organism 27

Introduction

Biology is the study of life. Through the study of biology students employ the processes of science in their investigations and explore the diversity of life and the inter-relationship between organisms and their environment. Students develop an understanding and knowledge of the unit of life – the cell – whose structures and processes are shared by all living organisms and, in so doing, gain an insight into the uniqueness, function and role of organisms, including themselves. In addition, they become aware of the use by humans of other living organisms and their products to enhance human health and the human environment and make informed evaluations about contemporary biological issues

It is intended that this syllabus will prove relevant to the lives of students and inspire in them an interest in and excitement about biology. It should enable them as future citizens to discuss and make judgements on issues in biology and science that impact on their daily lives and on society. It should provide them with the knowledge, skills and understanding to pursue further education, training and employment in biology-related fields and thereby respond to the needs of the economy and contribute to sustained economic development.

Aims

The aims of the syllabus are:
- to contribute to students' general education through their involvement in the process of scientific investigation and the acquisition of biological knowledge and understanding
- to encourage in students an attitude of scientific enquiry, of curiosity and self-discovery through
  (i)   individual study and personal initiative
  (ii) team work
  (iii) class-directed work
- to develop an understanding of biological facts and principles
- to enhance an interest in and develop an appreciation of the nature and diversity of organisms
- to create an awareness of the application of biological knowledge to modern society in personal, social, economic, environmental, industrial, agricultural, medical, waste management and other technological contexts
- to develop in students an ability to make informed evaluations about contemporary biological issues.

Syllabus Structure

The syllabus is composed of science for the enquiring mind or pure science, which constitutes approximately 70% of the syllabus, and the technological, political, social and economic aspects of biology, which constitutes the remaining 30%.

The syllabus consists of three units:

Unit One: Biology - The Study of Life
Unit Two: The Cell
Unit Three: The Organism.

The learning outcomes associated with the units of study are presented in four columns:

- Sub-units and topics
- Depth of treatment
- Contemporary issues and technology
- Practical activities.

The sequence in which the syllabus is presented does not imply any particular order of teaching. Teaching strategies should promote the aims and objectives of the syllabus. Professional discretion should be evident when dealing with sensitive topics in the syllabus.

Duration

The syllabus is designed for approximately 180 hours of class contact time (the equivalent of 270 class periods of 40 minutes duration or five class periods per week, to include at least one double period). A specific number of class periods for each sub-unit of the syllabus are recommended. These should be treated as a guideline intended to indicate the approximate amount of time needed. Teachers are encouraged to exercise discretion when allocating time periods to the various elements of the syllabus.

Practical Activities

In the course of their studies, students should undertake a range of practical work, laboratory work and fieldwork. Students should carry out these activities over the duration of the course. A record of this work should be retained.

In all practical work safety must be a major concern. Teachers are encouraged to develop in their students positive attitudes and approaches to safety in the range of activities they encounter and to inculcate in them an awareness of the values of creating a safe working environment.

Standard laboratory safety precautions should be observed and care taken when carrying out activities. All legal and health regulations must be adhered to in activities involving live and dead organisms. Before rearing and maintaining organisms, detailed information on the appropriate methods for the rearing and maintenance of the organisms must be studied. These methods must be strictly adhered to during the activity.

Students should appreciate the possibility for errors in activities and the precautions or controls that can be applied to reduce errors. Students should also be aware that the value of scientific method is limited by the extent of our own basic knowledge, by the basis of investigation, by our ability to interpret results, by its application to the natural world (which is always subject to change or variation) and by accidental discoveries.

By itemising activities, the syllabus aims to ensure that students attain certain skills including:

- manipulation of apparatus
- following instructions
- observation
- recording
- interpretation of observations and results
- practical enquiry and application of results.

Students should be encouraged to integrate information and communication technologies (ICTs) in their study of Biology.

Differentiation between Ordinary Level and Higher Level

Ordinary level and Higher level are differentiated on the basis of:

(i) **Range of topics:** The Higher level incorporates the Ordinary level. At Higher level an extended range of topics is required.

(ii) **Depth of treatment:** The Ordinary level course provides an overview of biology and its application to everyday life. At Higher level a deeper and more quantitative treatment of biology is required.

Where required, specific Higher level material is indicated at the end of each sub-unit as a Higher level Extension. This material has been printed in black throughout the syllabus.

The orientation of the Ordinary level course towards a more concrete and applied approach to study is enhanced by the inclusion of non-prescriptive material in the Teacher Guidelines, which accompanies the syllabus. At Ordinary level, the equivalent of 45 class periods has been allocated to this non-prescriptive material, which is distributed throughout the syllabus.

Assessment

The syllabus will be assessed in relation to its learning objectives through a terminal examination paper. All material within the syllabus is examinable. Practical work is an integral part of the study of biology. A practical assessment component may be introduced as part of the overall assessment at a later stage.

Objectives

The objectives of the syllabus are:

(a) Knowledge, Understanding and Skills

Students should have a knowledge and understanding of biological facts, terms, principles, concepts, relationships and experimental techniques, including practical laboratory skills.

Such skills should include

- an ability to carry out practical work, laboratory work and fieldwork activities safely and effectively
- an ability to record and interpret biological data.

(b) Application and Interface with Technology

Students should be able to apply, where possible, their knowledge and understanding of biology in environmental, industrial, agricultural, medical, waste management and other technological contexts.

(c) Science in the Political, Social and Economic Spheres

Students should be able to apply, where possible, their knowledge and understanding of biology in personal, social and economic spheres and to make informed evaluations about contemporary biological issues.

Leaving Certificate Biology

Unit One:
Biology - The Study of Life

UNIT ONE: BIOLOGY - THE STUDY OF LIFE

Sub-unit 1.1: The Scientific Method
Sub-unit 1.2: The Characteristics of Life
Sub-unit 1.3: Nutrition
Sub-unit 1.4: General Principles of Ecology
Sub-unit 1.5: A Study of an Ecosystem

Suggested Time Allowance in Class Periods:	Ordinary Level	Higher Level
Sub-unit 1.1: The Scientific Method	2	2
Sub-unit 1.2: The Characteristics of Life	3	3
Sub-unit 1.3: Nutrition	11	11
Sub-unit 1.4: General Principles of Ecology	8	13
Sub-unit 1.5: A Study of an Ecosystem	11	11
TOTAL	**35**	**40**

1.1 THE SCIENTIFIC METHOD

Sub-unit and Topic	Depth of Treatment	Contemporary Issues and Technology	Practical Activities
1.1.1 Biology	Definition and examples of the areas of study incorporated in biology.		
1.1.2 Scientific Method	Process of the scientific method.	Limitations of the scientific method.	The process of the scientific method should be developed as much as possible in all activities throughout the syllabus.
1.1.3 Experimentation	Principles of experimentation.		

1.2 THE CHARACTERISTICS OF LIFE

Sub-unit and Topic	Depth of Treatment	Contemporary Issues and Technology	Practical Activities
1.2.1 A Search for a Definition of Life	General outline of the diversity of living organisms. The common features and behaviours that identify them with the term "living". Definition of the terms "metabolism" and "continuity of life".		
1.2.2 Definition of Life	Definition of the term "life".		
1.2.3 Characteristics of Life	Definition and identification of the "characteristics of life" through the fundamental principles and interactions of organisation, nutrition, excretion, response, and reproduction.		

1.3 NUTRITION

Sub-unit and Topic	Depth of Treatment	Contemporary Issues and Technology	Practical Activities
1.3.1 Function of Food	Explanation, in simple terms, of the need for food.		
1.3.2 Chemical Elements	Identification of the elements present in food: six common elements, (C, H, N, O, P, S), five elements present in dissolved salts (Na, Mg, Cl, K, Ca) and three trace elements (Fe, Cu, Zn).		
1.3.3 Biomolecular Structures	Combination of elements in different ratios to form simple biomolecular units, e.g. carbohydrates $C_x(H_2O)_y$.		
1.3.4 Biomolecular Sources and the Components of Food	Carbohydrate, fat and oil (lipid), protein and vitamin: their basic element components, biomolecular components and sources. Vitamins: one water-soluble and one fat-soluble vitamin.		Conduct a qualitative test for: • starch • fat • a reducing sugar • a protein.
1.3.5 Energy Transfer Reactions	Definition of "anabolic" and "catabolic" reaction pathways. Photosynthesis as an example of an anabolic reaction sequence. Respiration as an example of a catabolic reaction sequence.		
1.3.6 Structural Role of Biomolecules	Carbohydrate — e.g. cellulose as a component of cell walls. Protein — e.g. fibrous proteins — as keratin in hair and skin, myosin in muscles. Lipid, e.g. component of cell membranes.		

1.3 NUTRITION (CONTINUED)			
Sub-unit and Topic	Depth of Treatment	Contemporary Issues and Technology	Practical Activities
1.3.7 Metabolic Role of Biomolecules	Carbohydrates and lipids as primary sources of energy for metabolic activity. Proteins as enzymes. Hormones as regulators of metabolic activity. Vitamins — e.g. C and D for tissue growth, cell production and health maintenance. Disorders associated with deficiency of a water-soluble and a fat-soluble vitamin.		
1.3.8 Minerals	Requirement and use of any two minerals present in dissolved salts or in trace amounts in: • plants • animals.		
1.3.9 Water	Importance of water for organisms.		

1.4 GENERAL PRINCIPLES OF ECOLOGY

Sub-unit and Topic	Depth of Treatment	Contemporary Issues and Technology	Practical Activities
1.4.1 Ecology	Definition of "ecology".		
1.4.2 Ecosystem	Definition and diversity of "ecosystems".		
1.4.3 Biosphere	Explanation of the term "biosphere".		
1.4.4 Habitat	Definition of "habitat".		
1.4.5 Environmental Factors	Definition and examples of the following as applied to terrestrial and aquatic environments: • abiotic factors • biotic factors • climatic factors. Definition and examples of edaphic factors as applied to terrestrial environments.		
1.4.6 Energy Flow	The sun as the primary source of energy for our planet. Feeding as a pathway of energy flow. Development of grazing food chain, food web and pyramid of numbers (explanation, construction, and use).		
1.4.7 Niche	Explanation of the term "niche".		
1.4.8 Nutrient Recycling	Nutrient recycling by organisms: definition. Outline of the Carbon Cycle and the Nitrogen Cycle. (Names of micro-organisms are not required).		
1.4.9 Human Impact on an Ecosystem	"Pollution" – definition, areas of effect, its control. Study the effects of any one pollutant. Definition of "conservation". "Waste management" – problems associated with waste disposal. Importance of waste minimisation.	Pollution: the ecological impact of one human activity. Outline of any one practice from one of the following areas: agriculture, fisheries, or forestry. Role of micro-organisms in waste management and pollution control.	

1.4 GENERAL PRINCIPLES OF ECOLOGY (CONTINUED)

Sub-unit and Topic	Depth of Treatment	Contemporary Issues and Technology	Practical Activities
H.1.4.10 Pyramid of Numbers (Extended Study)	Limitation of use. Inference of pyramid shape.		
H.1.4.11 Ecological Relationships	Factors that control populations. Definition and one example of the following control factors: • competition • predation • parasitism • symbiosis.		
H.1.4.12 Population Dynamics	Outline of the contributory factors or variables in predator and prey relationships.	The effect on the human population of: • war • famine • contraception • disease.	

1.5 A STUDY OF AN ECOSYSTEM

Sub-unit and Topic	Depth of Treatment	Contemporary Issues and Technology	Practical Activities
	Emphasis in this special study should be placed on the techniques of fieldwork and the recording and analysis of collected data.		
1.5.1 **Broad overview of a selected Ecosystem**	General overview of the diversity of life forms in an ecosystem.		Select and visit one ecosystem. Broad overview of the selected ecosystem.
1.5.2 **Observation and Scientific Study of a Selected Ecoystem**	Identification of a number of habitats from the selected ecosystem. Identification and application of collection apparatus available for an ecological study.		Identify any five fauna and any five flora using simple keys. Identify a variety of habitats within the selected ecosystem. Identify and use various apparatus required for collection methods in an ecological study.
1.5.3 **Organism Distribution**	Distinction between qualitative and quantitative surveys of a selected ecosystem for plants and animals. Familiarisation with frequency and percentage cover techniques available.		Conduct a quantitative study of plants and animals of a sample area of the selected ecosystem. Transfer results to tables, diagrams, graphs, histograms, or any other relevant mode. Identify possible sources of error in such a study.
1.5.4 **Choice of Habitat**	Relationship between an organism's suitability to its habitat and abiotic factors, to include any three of the following: pH, temperature (air and ground or aquatic), light intensity, water current, air current, dissolved oxygen, mineral content, percentage air in soil, percentage water in soil, percentage humus, salinity, degree of exposure, and slope.		Investigate any three abiotic factors present in the selected ecosystem, as listed. Relate results to choice of habitat selected by each organism identified in this study.

1.5 A STUDY OF AN ECOSYSTEM (CONTINUED)

Sub-unit and Topic	Depth of Treatment	Contemporary Issues and Technology	Practical Activities
1.5.5 Organism Adaptations	Necessity for structural, competitive or behavioural adaptation by organisms.		Note an adaptation feature by any organism in the selected ecosystem.
1.5.6 Organism Role in Energy Transfer	Identification of the role of the organism in energy transfer.		From the information obtained in this study construct food chains, a food web, and a pyramid of numbers.
1.5.7 Analysis	Necessity for analysis and assessment of results obtained.	Identification of local ecological issues related to the selected ecosystem.	Prepare a brief report of the results obtained.

Leaving Certificate Biology

UNIT TWO:
THE CELL

UNIT TWO: THE CELL

Sub-unit 2.1: Cell Structure
Sub-unit 2.2: Cell Metabolism
Sub-unit 2.3: Cell Continuity
Sub-unit 2.4: Cell Diversity
Sub-unit 2.5: Genetics

Suggested Time Allowance in Class Periods:	Ordinary Level	Higher Level
Sub-unit 2.1: Cell Structure	9	9
Sub-unit 2.2: Cell Metabolism	24	32
Sub-unit 2.3: Cell Continuity	3	4
Sub-unit 2.4: Cell Diversity	3	3
Sub-unit 2.5: Genetics	27	36
TOTAL	**66**	**84**

2.1 CELL STRUCTURE

Sub-unit and Topic	Depth of Treatment	Contemporary Issues and Technology	Practical Activities
2.1.1 **Microscopy**	An introduction to the microscope. Specific reference to the light microscope and the transmission electron microscope.		Be familiar with and use the light microscope.
2.1.2 **Cell Structure and Function**	Components of the cell as seen under the light microscope and their functions. Plant cells: cell wall, cytoplasm, nucleus, vacuole, and chloroplast. Animal cells: cytoplasm and nucleus. In both cases indicate the position and function of the cell membrane.		Prepare and examine one animal cell and one plant cell (e.g. own cheek cells, onion cells, *Elodea* leaf, potato tissue and moss) unstained and stained using the light microscope (x100, x400).
2.1.3 **Cell Ultrastructure**	Identification and function of the cell membrane, mitochondrion, chloroplast, nucleus, nuclear pores, ribosome, and DNA.		
H.2.1.4 **Prokaryotic and Eukaryotic Cells**	Existence and definition of "prokaryotic" and "eukaryotic" cells.		

2.2 CELL METABOLISM

Sub-unit and Topic	Depth of Treatment	Contemporary Issues and Technology	Practical Activities
2.2.1 Cell Metabolism	Definition of "metabolism".		
2.2.2 Sources of Energy	Reference to solar energy and cellular energy.		
2.2.3 Enzymes	Definition of "enzymes"—reference to their protein nature, folded shape, and roles in plants and animals. Special reference to their role in metabolism.		Investigate the effect of pH on the rate of one of the following: amylase, pepsin or catalase activity.
	Effect of pH and temperature on enzyme activity.		Investigate the effect of temperature on the rate of one of the following: amylase, pepsin or catalase activity.
		Bioprocessing with immobilised enzymes — procedure, advantages, and use in bioreactors.	Prepare one enzyme immobilisation and examine its application.
2.2.4 Photosynthesis	Definition and role of "photosynthesis". Representation by a balanced equation of the overall sequence of reactions.		
	A simple treatment of photosynthesis. Chlorophyll in chloroplasts traps sunlight energy. This trapped energy splits water to release electrons, protons, and oxygen. These electrons are passed to chlorophyll, the protons are released to a general pool of protons. The oxygen is either released to the atmosphere or used within the cell.		
	Electrons from chlorophyll are used with protons from the pool of protons and carbon dioxide to form a carbohydrate $C_x(H_2O)_y$.		
	Location of chlorophyll within cells.		
	Identification of the source of light, carbon dioxide and water for photosynthesis in leaf cells.	Human intervention: use of artificial light and carbon dioxide enrichment to promote crop growth in greenhouses.	Investigate the influence of light intensity or carbon dioxide on the rate of photosynthesis.

2.2 CELL METABOLISM (CONTINUED)

Sub-unit and Topic	Depth of Treatment	Contemporary Issues and Technology	Practical Activities
2.2.5 Respiration	Definition and role of "aerobic respiration". Representation by a balanced equation of the overall sequence of reactions for glucose. A simple treatment of aerobic respiration of glucose by reference to a two-stage process. Stage 1 does not require oxygen and releases a small amount of energy. Stage 2 does require oxygen and releases a large amount of energy. Definition of "anaerobic respiration". Reference to fermentation. Cellular location of the first and second-stage process.	Examine the role of micro-organisms in industrial fermentation, including bioprocessing with immobilised cells: procedure, advantages, and use in bioreactors.	Prepare and show the production of alcohol by yeast.
2.2.6 Movement through Cell Membranes	Selective permeability of membranes surrounding the cells and within the cells. Definition of the terms "diffusion" and "osmosis". Examples of each. Definition of "turgor". Simple explanation of turgidity in plant cells.	Describe the application of high salt or sugar concentration in food preservation.	Conduct any activity to demonstrate osmosis.
H.2.2.7 Enzymes (Extended Study)	The Active Site Theory to explain enzyme function and "specificity". Explanation of the term "optimum activity" under specific conditions as applied to pH range. Heat denaturation of protein.		Investigate the effect of heat denaturation on the activity of one enzyme.
H.2.2.8 Role of Adenosine Triphosphate (ATP) and Nicotinamide Adenine Dinucleotide (NAD)	Nature and role of ATP, production of ATP from ADP + ⓟ + Energy. Role of $NADP^+$ in trapping and transferring electrons and hydrogen ions in cell activities.		

2.2 CELL METABOLISM (CONTINUED)

Sub-unit and Topic	Depth of Treatment	Contemporary Issues and Technology	Practical Activities
H.2.2.9 Photosynthesis (Extended Study)	Photosynthesis as a two-stage process. The first stage, driven by light energy, is called the light stage or light-dependent stage. The second stage, which is dependent on the products of the light stage and does not require light, is called the dark stage or light-independent stage. In the light stage refer to the transfer of energy, the production of energised electrons, and their subsequent two-pathway systems. Pathway 1: by direct return to chlorophyll and the formation of ATP. Pathway 2: trapped by the reduction of $NADP^+$ to $NADP^-$; photolysis of water producing protons and electrons and releasing oxygen. H^+ is attracted to $NADP^-$ to form NADPH. In the dark stage, protons and electrons are transferred from NADPH to CO_2 in the production of $C_x(H_2O)_y$. Role of ATP. Regeneration of ADP and $NADP^+$ to the light stage. (Further biochemical references not required).		
H.2.2.10 Respiration (Extended Study)	First-stage process: Glycolysis — the conversion of a six-carbon carbohydrate to pyruvate with the generation of ATP. Fermentation option — ethanol or lactic acid production. Second-stage process: Production of Acetyl Co. A and one molecule of carbon dioxide. Krebs Cycle and the electron transport system, which produce more carbon dioxide, water, and ATP molecules. (Further biochemical references not required).		

2.3 CELL CONTINUITY

Sub-unit and Topic	Depth of Treatment	Contemporary Issues and Technology	Practical Activities
2.3.1 **Cell Continuity and Chromosome**	Explanation of the terms "cell continuity" and "chromosome".		
2.3.2 **Haploid, Diploid**	Definition of "haploid" and "diploid" number.		
2.3.3 **The Cell Cycle**	Description of cell activities in the state of non-division (interphase) and division (mitosis).	Cancer – definition and two possible causes.	
2.3.4 **Mitosis**	Definition of "mitosis". Simple treatment, with the aid of diagrams. (Names of stages and of chromosome parts are not required).		
2.3.5 **Function of Mitosis**	Primary function in single-celled and multicellular organisms.		
2.3.6 **Meiosis**	Definition of "meiosis".		
2.3.7 **Functions of Meiosis**	Functions of "meiosis".		
H.2.3.8 **Stages of Mitosis (Extended Study)**	Detailed study, with the aid of diagrams, of the stages of mitosis.		

2.4 CELL DIVERSITY			
Sub-unit and Topic	Depth of Treatment	Contemporary Issues and Technology	Practical Activities
2.4.1 Tissues	Definition of a "tissue". Exemplify by using four tissue types, two each from a plant and an animal.	Tissue culture: explanation and reference to any two applications.	
2.4.2 Organs	Definition of an "organ". Exemplify by using two kinds of organs, one each from a plant and an animal.		
2.4.3 Organ System	Definition of an "organ system". Exemplify by using any two animal organ systems.		

LEAVING CERTIFICATE BIOLOGY SYLLABUS

2.5 GENETICS

Sub-unit and Topic	Depth of Treatment	Contemporary Issues and Technology	Practical Activities
2.5.1 Variation of Species	Knowledge of the diversity of organisms. Definition of "species".		
2.5.2 Heredity and Gene Expression	Definition and example of "heredity" and "gene expression".		
2.5.3 Genetic Code	Definition and role of a "gene". Chromosome structure.		
2.5.4 DNA Structure, Replication and Profiling	Simple structure of DNA: two strands with Adenine(A)—Thymine(T), Guanine(G)—Cytosine(C) complement. Coding and non-coding structures. RNA as a complementary structure to DNA. Reference to uracil. Knowledge of the function of messenger RNA (mRNA).	DNA profiling: definition, any two applications. Stages involved: • cells are broken down to release DNA • DNA strands are cut into fragments using enzymes • fragments are separated on the basis of size • the pattern of fragment distribution is analysed.	Isolate DNA from a plant tissue.
	Replication of DNA involving the opening of the helix followed by the synthesis of complementary nucleic acid chains alongside the existing chains to form two identical helices.	Genetic screening: screening test diagnosis because of changed genes. (Detail of process not required).	
2.5.5 Protein Synthesis	Protein synthesis as follows: • DNA contains the code for protein • this code is transcribed to mRNA • the transcribed code goes to a ribosome • the code is translated and the amino acids are assembled in the correct sequence to synthesise the protein • the protein folds into its functional shape.		

2.5 GENETICS (CONTINUED)

Sub-unit and Topic	Depth of Treatment	Contemporary Issues and Technology	Practical Activities
2.5.6 Genetic Inheritance	Gamete formation. Definition of a "gamete" and its function in sexual reproduction in plants and animals. Definition of the following terms: • fertilisation • allele • homozygous and heterozygous • genotype • phenotype • dominance • recessive • incomplete dominance. Study of the inheritance to the first filial (F1) generation of a single unlinked trait in a cross involving: • homozygous parents • heterozygous parents • sex determination. The genotypes of parents, gametes and offspring should be shown.		
2.5.7 Causes of Variation	Variation from: sexual reproduction and mutations.	Two agents responsible for increased mutation rates.	
2.5.8 Evolution	Definition of "evolution". Theory of Natural Selection. Evidence from any one source.		
2.5.9 Genetic Engineering	Manipulation and alteration of genes. Process involving isolation, transformation, and expression.	Three applications: one plant, one animal, one micro-organism.	

2.5 GENETICS (CONTINUED)

Sub-unit and Topic	Depth of Treatment	Contemporary Issues and Technology	Practical Activities
H.2.5.10 Origin of the Science of Genetics	Work of Gregor Mendel leading to the expression of his findings in two laws.		
H.2.5.11 Law of Segregation	State and explain the Law of Segregation.		
H.2.5.12 Law of Independent Assortment	State and explain the Law of Independent Assortment.		
H.2.5.13 Dihybrid Cross	Study of the inheritance to the second filial generation (F2) of two unlinked traits using the Punnett square technique. Definition of linkage. Explanation of change in 1:1:1:1 probability for a dihybrid heterozygote crossed with a dihybrid recessive organism. (Knowledge of crossing over is not required). Sex linkage. Non-nuclear inheritance, e.g. mitochondrial and chloroplast DNA.		
H.2.5.14 Nucleic Acid Structure and Function (Extended Study)	DNA structure, to include: deoxyribose sugar, phosphate, and four named nitrogenous bases. Specific purine and pyrimidine couples — complementary base pairs. Hydrogen bonding. Double helix.		
H.2.5.15 Protein Synthesis (Extended Study)	Location of protein synthesis, process of protein synthesis — reference only to molecular involvement of DNA, mRNA, tRNA, rRNA and amino acids to provide an understanding of their role in coding information.		

LEAVING CERTIFICATE BIOLOGY

UNIT THREE: THE ORGANISM

LEAVING CERTIFICATE BIOLOGY SYLLABUS

UNIT THREE: THE ORGANISM

Sub-unit 3.1 Diversity of Organisms
Sub-unit 3.2 Organisation and the Vascular Structures
Sub-unit 3.3 Transport and Nutrition
Sub-unit 3.4 Breathing System and Excretion
Sub-unit 3.5 Responses to Stimuli
Sub-unit 3.6 Reproduction and Growth

Suggested Time Allowance in Class Periods:	Ordinary Level	Higher Level
Sub-unit 3.1: Diversity of Organisms	14	17
Sub-unti 3.2: Organisation and the Vascular Structures	21	24
Sub-unit 3.3: Transport and Nutrition	15	16
Sub-unit 3.4: Breathing System and Excretion	12	14
Sub-unit 3.5: Responses to Stimuli	32	37
Sub-unit 3.6: Reproduction and Growth	30	38
TOTAL	124	146

3.1 DIVERSITY OF ORGANISMS

Sub-unit and Topic	Depth of Treatment	Contemporary Issues and Technology	Practical Activities
3.1.1 Diversity of Organisms	Five-kingdom system of classification: Monera (Prokaryotae), Protista (Protoctista), Fungi, plant, and animal. (Further sub-classification not required).		
3.1.2 Micro-organisms	Distribution of bacteria and fungi in nature.		
3.1.3 Monera, e.g. Bacteria	Bacterial cells: basic structure (including plasmid DNA), three main types. Reproduction. Nutrition.		
	Factors affecting growth.	Economic importance of bacteria: examples of any two beneficial and any two harmful bacteria.	
	Understanding of the term "pathogenic".		
	Definition and role of "antibiotics".	Potential abuse of antibiotics in medicine.	
3.1.4 Fungi	Saprophytic and parasitic forms.	Mention of edible and poisonous fungi.	
	Rhizopus: structure and life cycle.		
	Nutrition.		
	Yeast: structure and reproduction (budding).	Economic importance of fungi: examples of any two beneficial and any two harmful fungi.	Investigate the growth of leaf yeast using agar plates and controls.
3.1.5 Laboratory Procedures when handling Micro-organisms	Precautions when working with micro-organisms. Asepsis and sterility: definition of each term as applied to living organisms.		
	Containment and disposal.		
3.1.6 Protista, e.g. Amoeba	*Amoeba* — cell organisation to include nucleus and sub-cellular structures.		
3.1.7 Plant, e.g. the Flowering Plant	Plant kingdom as exemplified by the flowering plant. (Refer to the remaining Sub-units of Unit 3).		

3.1 DIVERSITY OF ORGANISMS (CONTINUED)

Sub-unit and Topic	Depth of Treatment	Contemporary Issues and Technology	Practical Activities
3.1.8 Animal, e.g. the Human	Animal kingdom as exemplified by the human. (Refer to the remaining Sub-units of Unit 3.)		
H.3.1.9 Nature of Bacteria and Fungi	Prokaryotic nature of bacteria. Eukaryotic nature of fungi.		
H.3.1.10 Growth Curves	Growth curves of micro-organisms.	Batch and continuous flow food processing.	

3.2 ORGANISATION AND THE VASCULAR STRUCTURES

Sub-unit and Topic	Depth of Treatment	Contemporary Issues and Technology	Practical Activities
3.2.1 Organisational Complexity of the Flowering Plant	Organisational complexity of the flowering plant as exemplified by the root, stem, leaf, flower, seed, and transport/vascular structures. Function of the root and shoot system. Explanation of the term "meristem" — location in the root and shoot. Location of three tissue types — dermal, ground and vascular in transverse and in longitudinal sections of the root and stem. Xylem and phloem as examples of vascular tissues — their function and structure. Identification of dicotyledons and monocotyledons under the headings: woody/herbaceous, arrangement of floral parts, arrangement of vascular bundles, cotyledon or seed leaf number.		Prepare and examine microscopically the transverse section of a dicotyledonous stem (x100, x400).
3.2.2 Organisational Complexity of the Human	Organisational complexity of the human. The circulatory system: description of the structures and organisation of tissues in the closed circulatory system in humans, strong muscular heart and vessels (arteries, veins, capillaries, venules, arterioles). Role of muscle tissues and valves. Two-circuit circulatory system. Drawing of the structure of the heart, the main pathways of blood circulation, including the hepatic portal system. Cardiac supply through the cardiac artery and vein. Simple understanding of: • heartbeat and its control • pulse • blood pressure.	Knowledge of the effect of smoking, diet and exercise on the circulatory system.	Dissect, display and identify an ox's or a sheep's heart. Investigate the effect of exercise on the breathing rate or pulse of a human.

3.2 ORGANISATION AND THE VASCULAR STRUCTURES (CONTINUED)

Sub-unit and Topic	Depth of Treatment	Contemporary Issues and Technology	Practical Activities
	The lymphatic system: • structure: lymph nodes, lymph vessels • any three functions. Composition of blood, role of red blood cells, white blood cells, platelets, and plasma. (Classification of white blood cells not required). Blood grouping — names of the common blood groups A, B, AB, O and the Rhesus factors.		
H.3.2.3 Blood Cells (Extended Study)	More detailed treatment of red blood cells — e.g. absence of nucleus, absence of mitochondria. White blood cells — classification as lymphocytes and monocytes.		
H.3.2.4 Heartbeat Control	An awareness of specialised heart muscle tissue and the existence and location of pacemaker nodes (SA and AV). The heart cycle, systole and diastole periods.		

3.3 TRANSPORT AND NUTRITION

Sub-unit and Topic	Depth of Treatment	Contemporary Issues and Technology	Practical Activities
3.3.1 Nutrition in the Flowering Plant	Autotrophic nature of plants. Description of the uptake and process of transport of the following through the plant: • water: to include reference to root hairs, root cortex, xylem, osmosis, diffusion, root pressure, transpiration, and stomata • minerals: to include solubility in water, transport from the roots to all parts of the plant by the same route as water • carbon dioxide: directly from respiring cells or through stomata • photosynthetic products: production of carbohydrate and transport through phloem sieve tube cells.		
3.3.2 Modified Plant Food Storage Organs	One example of a root, stem and leaf modification as a food storage organ.		
3.3.3 Nutrition in the Human	Heterotrophic organisms — "omnivore" (human), "herbivore" and "carnivore" — definition of terms. Explanation of the term "digestion". Outline the need for digestion and a digestive system. Explanation of the terms "ingestion", "digestion", "absorption" and "egestion" as related to the sequence in the human digestive tract.		

3.3 TRANSPORT AND NUTRITION (CONTINUED)			
Sub-unit and Topic	Depth of Treatment	Contemporary Issues and Technology	Practical Activities
3.3.4 Human Digestive System	Macrostructure and basic function of the alimentary canal and associated glands in the digestion and transport of nutrients. Explanation of the mechanical breakdown and transport of food, to include the role of teeth, peristalsis, and the stomach. Explanation of the chemical breakdown of food, to include: • bile salts • the role, production site, pH at a named location of action and products of an amylase, a protease and a lipase enzyme. Two functions of symbiotic bacteria in the digestive tract. Benefits of fibre. Basic structure of the small intestine and large intestine in relation to their functions.		
3.3.5 Blood Transport of Nutrients	Description of the composition of blood fluid as a transport system of nutrients, the absorption of nutrients from the villi, transport through the hepatic portal vein to the liver. The function of the liver (without biochemical pathways). The transport of nutrients to all nutrient-requiring cells of the body, and the transport of waste products to the kidney.		
3.3.6 Balanced Human Diet	Explain the concept of a balanced diet, variety, and moderation. Relate its importance to age, sex and activity (detailed breakdown not required), and to variety from a selection of food groups — milk and milk products; meat, fish and poultry; breads and cereals; fruit and vegetables; others, e.g. fats, oils, alcohol.		
H.3.3.7 Cohesion – Tension Model of Xylem Transport	As related to attractive forces of water molecules, cohesive property, role of transpiration. Refer to the work of Dixon and Joly.		

3.4 BREATHING SYSTEM AND EXCRETION

Sub-unit and Topic	Depth of Treatment	Contemporary Issues and Technology	Practical Activities
3.4.1 Homeostasis	Definition of "homeostasis".		
3.4.2 Necessity for Homeostasis	The necessity for homeostasis in living organisms.		
3.4.3 The Structure of an Exchange System in Flowering Plants	Examination of the structure of the leaf in relation to gaseous exchange. Reference to the presence of lenticels in stem structures.		
3.4.4 The Breathing System in the Human	Macrostructure and basic function of the breathing tract in humans. Essential features of the alveoli and capillaries as surfaces over which gas exchange takes place. Description of the mechanism of the breathing system in the exchange of gases in humans.	Breathing disorders: one example of a breathing disorder, from the following: asthma and bronchitis; one possible cause, prevention, and treatment.	
3.4.5 Plant Excretion	The role of leaves as excretory organs of plants.		
3.4.6 The Excretory System in the Human	Role of the excretory system in homeostasis. Function, location and excretory products of the lungs, skin, and urinary system. Macrostructure and basis function of the urinary excretory system in humans (kidney, ureters, urinary bladder, and urethra). Role of the kidney in regulating body fluids. Identification of the site of filtration. Re-absorption in the cortex, in the medulla and renal pelvis. Description of the pathway of urine from the kidney to the urethra.		

3.4 BREATHING SYSTEM AND EXCRETION (CONTINUED)

Sub-unit and Topic	Depth of Treatment	Contemporary Issues and Technology	Practical Activities
H.3.4.7 Carbon Dioxide: A Controlling Factor in Gaseous Exchange	Carbon dioxide level as a controlling factor in stomatal opening and in the human breathing (respiratory) system.		
H.3.4.8 The Nephron as a Unit of Kidney Function	The nephron structure and its associated blood supply. Formation of urine: Bowman's capsule, passage of glomerular filtrate through the proximal convoluted tubule, where reabsorption of required body substances takes place — glucose, amino acids, some salts and water reabsorbed into the blood by osmosis, diffusion, and active transport. More water reabsorbed in the Loop of Henle and the distal convoluted tube. Urine passes into the pelvis of the kidney and to the bladder for storage. Reabsorption of water in the collecting duct is under hormonal influence (ADH). Its action depends on the water content of the blood. (No further details required).		

3.5 RESPONSES TO STIMULI

Sub-unit and Topic	Depth of Treatment	Contemporary Issues and Technology	Practical Activities
3.5.1 Structures for Response	Chemical or hormonal system, nerve and sense organ system, muscular, skeletal and an immune system.		
3.5.2 Responses in the Flowering Plant	Growth regulation. Tropisms: definition of the following: "phototropism", "geotropism", "thigmatropism", "hydrotropism", and "chemotropism". Examples of phototropism and geotropism.		
	Regulatory system: definition of a "growth regulator", transport through the vascular system, combined effect, growth promoter and growth inhibitor. Name four methods of anatomical or chemical adaptation that protect plants.	Use of plant regulators: any two examples.	Investigate the effect of I.A.A. growth regulator on plant tissue.
3.5.3 Responses in the Human	The nervous system: two-part division into the central nervous system (CNS) and the peripheral nervous system (PNS). Neuron: its structure and function, with reference only to cell body, dendrites, axon, myelin sheath, Schwann cell, and neurotransmitter vesicles. Movement of nerve impulse. (Detailed knowledge of electro-chemistry not required). Synapse. Activation and inactivation of neurotransmitter. Role and position of three types of neuron: sensory, motor and interneuron. The senses, with the brain as an interpreting centre. Knowledge of the senses. Study of the eye and the ear. Corrective measures for long and short sight or for hearing.		

3.5 RESPONSES TO STIMULI (CONTINUED)			
Sub-unit and Topic	Depth of Treatment	Contemporary Issues and Technology	Practical Activities
	Note: The following are not required: biochemical action, detailed structure of cochlea and semicircular canals in ear, names of sensory receptors in the skin.		
	Central nervous system: brain and spinal cord. Location and function of the following parts of the brain: cerebrum, hypothalamus, pituitary gland, cerebellum, and medulla oblongata. Cross-section of spinal cord indicating: white matter, grey matter and central canal (refer to their constituent bodies), three-layer protective tissue — the meninges. Dorsal and ventral roots of the spinal nerve.	Nervous system disorders: any one example of a nervous system disorder, from the following: paralysis and Parkinson's disease; one possible cause, prevention, and treatment.	
	Peripheral nervous system: location of nerve fibres and cell bodies. Role, structure and mechanisms of the reflex action. (Cranial nerves, sympathetic and parasympathetic systems are not required).		
	Endocrine system: definition of a "hormone". Comparison with nerve action, distinction between exocrine and endocrine glands, with examples. Location of the principal endocrine glands in the human. For each of the glands name one hormone and give its functions. For one hormone give a description of its deficiency symptoms, excess symptoms, and corrective measures.	Hormone supplements: two examples of their use.	
	Musculoskeletal system: description of the structure and functions of the skeleton. Component parts of the axial skeleton: skull, vertebrae, ribs, and sternum. Position and function of discs in relation to vertebrae.		
	Component parts of the appendicular skeleton: pectoral and pelvic girdles and their attached limbs.		

3.5 RESPONSES TO STIMULI (CONTINUED)

Sub-unit and Topic	Depth of Treatment	Contemporary Issues and Technology	Practical Activities
	Macroscopic anatomy of a long bone: medullary cavity, compact bone, spongy bone, and cartilage.		
	Function of the following: cartilage, compact bone, spongy bone (include red and yellow marrows). (T.S of bone is not required).		
	Classification, location and function of joints: immovable, slightly movable, free-moving or synovial.	Disorders of the musculoskeletal system: one example of a musculoskeletal disorder, from the following: arthritis and osteoporosis; one possible cause, prevention, and treatment.	
	Role of cartilage and ligaments in joints.		
	Role of tendons.		
	General relation of muscles to the skeleton — antagonistic muscle pairs as exemplified by one human pair.		
	The defence system in humans: general defence system to include the skin and mucous membrane lining of the breathing, reproductive and digestive tracts. Phagocytic white blood cells.		
	Specific defence system (immune system): antigen antibody response. Definition of "induced immunity".	Vaccination and immunisation.	
3.5.4 Viruses	Viruses: identify the problem of definition. Variety of shapes. Basic structure. Viral reproduction.	Economic and medical importance of viruses: two harmful examples, one beneficial example.	

3.5 RESPONSES TO STIMULI (CONTINUED)

Sub-unit and Topic	Depth of Treatment	Contemporary Issues and Technology	Practical Activities
H.3.5.5 Auxins	Study auxin as an example of a plant growth regulator under the headings of production site(s), function, and effects.		
H.3.5.6 Plant Growth Regulators and Animal Hormones (Extended Study)	Explanation of the mechanism of plant response to any one external stimulus. Description of the feedback mechanism of any one animal hormonal system.		
H.3.5.7 Human Immune System (Extended Study)	Role of lymphocytes: B and T cell types. Role of B cells in antibody production. Role of T cells as helpers, killers, suppressors, and memory T cells.		
H.3.5.8 Growth and Development in Bones	Osteoblast role in bone growth. Terminating development of adult height. Role of osteoblasts in bone cell replacement. Bone renewal. Role of calcium in bone.		

3.6 REPRODUCTION AND GROWTH			
Sub-unit and Topic	Depth of Treatment	Contemporary Issues and Technology	Practical Activities
3.6.1 Reproduction of the Flowering Plant	Sexual: Structure and function of the floral parts: sepal, petal, stamen, and carpel. (Terms "calyx", "corolla", "androecium" and "gynoecium" not required). Pollen grain produces male gametes (statement only). Embryo sac produces an egg cell and polar nuclei (statement only). Definition and methods of "pollination": self-pollination and cross-pollination, to include wind and animal. Definition of "fertilisation": • fertilisation of an egg to form a diploid zygote, which develops into an embryo • second fertilisation with polar nuclei results in the formation of the endosperm. Seed structure and function of the following parts: Testa, plumule, radicle, embryo and cotyledon attachments. Embryo and a food supply as contained either in an endosperm or in seed leaves (the cotyledons). Monocotyledon, dicotyledon classification and distinguishing features. Reference to non-endospermic seed. Fruit formation — simple statement. (Classification of fruits not required). Fruit and seed dispersal: examples of wind, water, animal and self-dispersal. Emphasise the need for dispersal. Definition and advantages of "dormancy".	Seedless fruit production caused by genetic variety of plants and growth regulators. Mention of dormancy in agricultural and horticultural practices.	

3.6 REPRODUCTION AND GROWTH (CONTINUED)

Sub-unit and Topic	Depth of Treatment	Contemporary Issues and Technology	Practical Activities
	"Germination": definition, factors necessary, role of digestion and respiration. Stages of seedling growth.		Investigate the effect of water, oxygen and temperature on germination. Use starch agar or skimmed milk plates to show digestive activity during germination.
3.6.2 **Sexual Reproduction in the Human**	Vegetative propagation: asexual reproduction in plants. One example each from stem, root, leaf, and bud. Comparison of reproduction by seed and by vegetative propagation. General structure of the reproductive system — male and female. Functions of the main parts. Role of meiosis in the production of sperm cells and egg (ova). (The detailed treatment of spermatogenesis and oogenesis are not required). Definition of "secondary sexual characteristics". Role of oestrogen, progesterone, and testosterone. The menstrual cycle: the events and outlined role of oestrogen and progesterone.	Artificial propagation in flowering plants, any four methods used to artificially propagate plants.	

3.6 REPRODUCTION AND GROWTH (CONTINUED)			
Sub-unit and Topic	Depth of Treatment	Contemporary Issues and Technology	Practical Activities
	Copulation.	Birth control – natural, mechanical, chemical and surgical methods of contraception.	
	Location of fertilisation.	Infertility: One cause of male infertility from the following disorders: low sperm count, low sperm mobility, endocrine gland failure. Availability of corrective measures. One cause of female infertility from the following disorders: blockage of the Fallopian tube, endocrine gland failure. Availability of corrective measures.	
	Implantation, placenta formation and function. (Detailed embryological terms not required). Birth – outline of process. Milk production and breastfeeding.	In-vitro fertilisation and implantation. Biological benefits of breastfeeding.	

3.6 REPRODUCTION AND GROWTH (CONTINUED)

Sub-unit and Topic	Depth of Treatment	Contemporary Issues and Technology	Practical Activities
H.3.6.3 Sexual Reproduction in the Flowering Plant (Extended Study)	Pollen grain development from microspore mother cells: meiotic division, mitotic division, generative and tube nuclei production, formation of pollen grain. Embryo sac development: megaspore mother cell, meiotic division, cell disintegration, mitotic division in the production of eight cells of the embryo sac, one of which is the egg cell. (Antipodal cells and synergids not required).		
H.3.6.4 Human Embryo Development (Extended Study)	Sequence of development from fertilised egg, morula, blastocyst, existence of amnion, placenta formation from embryonic and uterine tissue. Development of embryo up to third month.		
H.3.6.5 Sexual Reproduction in the Human (Extended Study)	Detailed study of the menstrual cycle and hormonal control.	Menstrual disorders: one example of a menstrual disorder from the following: endometriosis and fibroids; one possible cause, prevention and treatment	

LEAVING CERTIFICATE BIOLOGY SYLLABUS